U.S. Department of Justice
Office of Justice Programs
National Institute of Justice

A GUIDE for
Explosion
and
Bombing
Scene Investigation

research report

U.S. Department of Justice
Office of Justice Programs
810 Seventh Street N.W.
Washington, DC 20531

Janet Reno
Attorney General

Daniel Marcus
Acting Associate Attorney General

Mary Lou Leary
Acting Assistant Attorney General

Julie E. Samuels
Acting Director, National Institute of Justice

Office of Justice Programs **World Wide Web Site** *http://www.ojp.usdoj.gov*	**National Institute of Justice** **World Wide Web Site** *http://www.ojp.usdoj.gov/nij*

About the National Institute of Justice

The National Institute of Justice (NIJ), a component of the Office of Justice Programs, is the research agency of the U.S. Department of Justice. Created by the Omnibus Crime Control and Safe Streets Act of 1968, as amended, NIJ is authorized to support research, evaluation, and demonstration programs, development of technology, and both national and international information dissemination. Specific mandates of the Act direct NIJ to:

- Sponsor special projects and research and development programs that will improve and strengthen the criminal justice system and reduce or prevent crime.

- Conduct national demonstration projects that employ innovative or promising approaches for improving criminal justice.

- Develop new technologies to fight crime and improve criminal justice.

- Evaluate the effectiveness of criminal justice programs and identify programs that promise to be successful if continued or repeated.

- Recommend actions that can be taken by Federal, State, and local governments as well as by private organizations to improve criminal justice.

- Carry out research on criminal behavior.

- Develop new methods of crime prevention and reduction of crime and delinquency.

In recent years, NIJ has greatly expanded its initiatives, the result of the Violent Crime Control and Law Enforcement Act of 1994 (the Crime Act), partnerships with other Federal agencies and private foundations, advances in technology, and a new international focus. Examples of these new initiatives include:

- Exploring key issues in community policing, violence against women, violence within the family, sentencing reforms, and specialized courts such as drug courts.

- Developing dual-use technologies to support national defense and local law enforcement needs.

- Establishing four regional National Law Enforcement and Corrections Technology Centers and a Border Research and Technology Center.

- Strengthening NIJ's links with the international community through participation in the United Nations network of criminological institutes, the U.N. Criminal Justice Information Network, and the NIJ International Center.

- Improving the online capability of NIJ's criminal justice information clearinghouse.

- Establishing the ADAM (Arrestee Drug Abuse Monitoring) program—formerly the Drug Use Forecasting (DUF) program—to increase the number of drug-testing sites and study drug-related crime.

The Institute Director establishes the Institute's objectives, guided by the priorities of the Office of Justice Programs, the Department of Justice, and the needs of the criminal justice field. The Institute actively solicits the views of criminal justice professionals and researchers in the continuing search for answers that inform public policymaking in crime and justice.

To find out more about the National Institute of Justice,
please contact:

National Criminal Justice Reference Service,
P.O. Box 6000
Rockville, MD 20849–6000
800–851–3420
e-mail: *askncjrs@ncjrs.org*

To obtain an electronic version of this document, access the NIJ Web site
(*http://www.ojp.usdoj.gov/nij/pubs-sum/181869.htm*).

If you have questions, call or e-mail NCJRS.

A Guide for Explosion and Bombing Scene Investigation

Written and Approved by the
Technical Working Group for Bombing Scene Investigation

June 2000
NCJ 181869

Julie E. Samuels
Acting Director

David G. Boyd, Ph.D.
Deputy Director

Richard M. Rau, Ph.D.
Project Monitor

The National Institute of Justice is a component of the Office of Justice Programs, which also includes the Bureau of Justice Assistance, the Bureau of Justice Statistics, the Office of Juvenile Justice and Delinquency Prevention, and the Office for Victims of Crime.

Message From the Attorney General

The investigation conducted at the scene of an explosion or bombing plays a vital role in uncovering the truth about the incident. The evidence recovered can be critical in identifying, charging, and ultimately convicting suspected criminals. For this reason, it is absolutely essential that the evidence be collected in a professional manner that will yield successful laboratory analyses. One way of ensuring that we, as investigators, obtain evidence of the highest quality and utility is to follow sound protocols in our investigations.

Recent cases in the criminal justice system have brought to light the need for heightened investigative practices at all crime scenes. In order to raise the standard of practice in explosion and bombing investigations of both small and large scale, in both rural and urban jurisdictions, the National Institute of Justice teamed with the National Center for Forensic Science at the University of Central Florida to initiate a national effort. Together they convened a technical working group of law enforcement and legal practitioners, bomb technicians and investigators, and forensic laboratory analysts to explore the development of improved procedures for the identification, collection, and preservation of evidence at explosion and bombing scenes.

This *Guide* was produced with the dedicated and enthusiastic participation of the seasoned professionals who served on the Technical Working Group for Bombing Scene Investigation. These 32 individuals brought together knowledge and practical experience from Federal law enforcement agencies—as well as from large and small jurisdictions across the United States—with expertise from national organizations and abroad. I applaud their efforts to work together over the course of 2 years in developing this consensus of recommended practices for public safety personnel.

In developing its investigative procedures, every jurisdiction should give careful consideration to those recommended in this *Guide* and to its own unique local conditions and logistical circumstances. Although factors that vary among investigations may call for different approaches or even preclude the use of certain procedures described in the *Guide,* consideration of the *Guide's* recommendations may be invaluable to a jurisdiction shaping its own protocols. As such, *A Guide for Explosion and Bombing Scene Investigation* is an important tool for refining investigative practices dealing with these incidents, as we continue our search for truth.

Janet Reno

Message From the President of the University of Central Florida

The University of Central Florida (UCF) is proud to take a leading role in the investigation of fire and explosion scenes through the establishment of the National Center for Forensic Science (NCFS). The work of the Center's faculty, staff, and students, in cooperation with the National Institute of Justice (NIJ), has helped produce the NIJ Research Report *A Guide for Explosion and Bombing Scene Investigation.*

More than 150 graduates of UCF's 25-year-old program in forensic science are now working in crime laboratories across the country. Our program enjoys an ongoing partnership with NIJ to increase knowledge and awareness of fire and explosion scene investigation. We anticipate that this type of mutually beneficial partnership between the university, the criminal justice system, and private industry will become even more prevalent in the future.

As the authors of this *Guide* indicate, the field of explosion and bombing investigation lacks nationally coordinated investigative protocols. NCFS recognizes the need for this coordination. The Center maintains and updates its training criteria and tools so that it may serve as a national resource for public safety personnel who may encounter an explosion or bombing scene in the line of duty.

I encourage interested and concerned public safety personnel to use *A Guide for Explosion and Bombing Scene Investigation.* The procedures recommended in the *Guide* can help to ensure that more investigations are successfully concluded through the proper identification, collection, and examination of all relevant forensic evidence.

Dr. John C. Hitt

Technical Working Group for Bombing Scene Investigation

The Technical Working Group for Bombing Scene Investigation (TWGBSI) is a multidisciplinary group of content area experts from the United States, Canada, and Israel, each representing his or her respective agency or practice. Each of these individuals is experienced in the investigation of explosions, the analysis of evidence gathered, or the use in the criminal justice system of information produced by the investigation. They represent such entities as fire departments, law enforcement agencies, forensic laboratories, private companies, and government agencies.

At the outset of the TWGBSI effort, the National Institute of Justice (NIJ) and the National Center for Forensic Science (NCFS) created the National Bombing Scene Planning Panel (NBSPP)—composed of distinguished law enforcement officers, representatives of private industry, and researchers—to define needs, develop initial strategies, and steer the larger group. Additional members of TWGBSI were then selected from recommendations solicited from NBSPP; NIJ's regional National Law Enforcement and Corrections Technology Centers; and national organizations and agencies such as the Federal Bureau of Investigation, the Bureau of Alcohol, Tobacco and Firearms, the American Society of Crime Laboratory Directors, and the National District Attorneys Association.

Collectively, over a 2-year period, the 32 members of TWGBSI listed below worked together to develop this handbook, *A Guide for Explosion and Bombing Scene Investigation.*

National Bombing Scene Planning Panel of TWGBSI

Joan K. Alexander
Office of the Chief State's
  Attorney
Rocky Hill, Connecticut

Roger E. Broadbent
Virginia State Police
Fairfax, Virginia

John A. Conkling, Ph.D.
American Pyrotechnics
  Association
Chestertown, Maryland

Sheldon Dickie
Royal Canadian Mounted
   Police
Gloucester, Ontario, Canada

Ronald L. Kelly
Federal Bureau of
   Investigation
Washington, D.C.

Jimmie C. Oxley, Ph.D.
University of Rhode Island
Kingston, Rhode Island

Roger N. Prescott
Austin Powder Company
Cleveland, Ohio

James C. Ronay
Institute of Makers of
   Explosives
Washington, D.C.

James T. Thurman
Eastern Kentucky University
Richmond, Kentucky

Carl Vasilko
Bureau of Alcohol, Tobacco
   and Firearms
Washington, D.C.

Raymond S. Voorhees
U.S. Postal Inspection Service
Dulles, Virginia

Technical Working Group for Bombing Scene Investigation

Andrew A. Apollony
Federal Bureau of
   Investigation
Quantico, Virginia

Michael Boxler
Bureau of Alcohol, Tobacco
   and Firearms
St. Paul, Minnesota

Steven G. Burmeister
Federal Bureau of
   Investigation
Washington, D.C.

Gregory A. Carl
Federal Bureau of
   Investigation
Washington, D.C.

Stuart W. Case
Forensic Consulting Services
Pellston, Michigan

Lance Connors
Hillsborough County Sheriff's
   Office
Tampa, Florida

James B. Crippin
Colorado Bureau of
   Investigation
Pueblo, Colorado

John E. Drugan
Massachusetts State Police
Sudbury, Massachusetts

Dirk Hedglin
Great Lakes Analytical, Inc.
St. Clair Shores, Michigan

Larry Henderson
Kentucky State Police
Lexington, Kentucky

Thomas H. Jourdan, Ph.D.
Federal Bureau of
   Investigation
Washington, D.C.

Frank Malter
Bureau of Alcohol, Tobacco
   and Firearms
Washington, D.C.

Thomas J. Mohnal
Federal Bureau of
   Investigation
Washington, D.C.

David S. Shatzer
Bureau of Alcohol, Tobacco
   and Firearms
Washington, D.C.

Patricia Dawn Sorenson
Naval Criminal Investigative
   Service
San Diego, California

Frank J. Tabert
International Association of
   Bomb Technicians and
   Investigators
Franklin Square, New York

Calvin K. Walbert
Chemical Safety and Hazard
   Investigation Board
Washington, D.C.

Leo W. West
Federal Bureau of
   Investigation
Washington, D.C.

Carrie Whitcomb
National Center for Forensic
   Science
Orlando, Florida

David M. Williams
Lockheed Martin Energy
   Systems
Oak Ridge, Tennessee

Jehuda Yinon, Ph.D.
Weizmann Institute of Science
Rehovot, Israel

Acknowledgments

The National Institute of Justice (NIJ) acknowledges, with great thanks, the members of the Technical Working Group for Bombing Scene Investigation (TWGBSI) for their extensive efforts on this project and their dedication to improving the level of explosion and bombing investigations for the good of the criminal justice system. Each of the 32 members of this network of experts gave their time and expertise to draft and review the *Guide,* providing feedback and perspective from a variety of disciplines and from all areas of the United States, Canada, and Israel. The true strength of this *Guide* is derived from their commitment to develop procedures that could be implemented across the country, from rural townships to metropolitan areas. In addition, thanks are extended to the agencies and organizations the Technical Working Group (TWG) members represent for their flexibility and support, which enabled the participants to see this project to completion.

NIJ is immensely grateful to the National Center for Forensic Science (NCFS) at the University of Central Florida, particularly Director Carrie Whitcomb and Project Coordinator Joan Jarvis, for its coordination of the TWGBSI effort. NCFS's support in planning and hosting the TWG meetings, as well as the support of its staff in developing the *Guide,* made this work possible.

Additionally, thanks are extended to all the individuals, agencies, and organizations across the country who participated in the review of this *Guide* and provided valuable comments and input. In particular, thanks go to the Bureau of Alcohol, Tobacco and Firearms, the Federal Bureau of Investigation, the National District Attorneys Association, the American Society of Crime Laboratory Directors, the International Association of Arson Investigators, and the International Association of Bomb Technicians and Investigators. While all review comments were given careful consideration by the TWG in developing the final document, the review by these organizations is not intended to imply their endorsement of the *Guide.*

NIJ would like to thank the co-manager for this project, Kathleen Higgins, for her advice and significant contribution to the development of the *Guide*.

Special thanks go to former NIJ Director Jeremy Travis for his support and guidance and to Lisa Forman, Lisa Kaas, and Anjali Swienton for their contributions to the TWG program. Thanks also go to Rita Premo of Aspen Systems Corporation, who provided tireless work editing and re-editing the various drafts of the *Guide*.

Finally, NIJ would like to acknowledge Attorney General Janet Reno, whose support and commitment to the improvement of the criminal justice system made this work possible.

Contents

Introduction

"I had imagined that Sherlock Holmes would have at once hurried into the house and plunged into a study of the mystery. Nothing appeared to be further from his intention. He lounged up and down the pavement and gazed vacantly at the ground, the sky, the opposite houses. Having finished his scrutiny, he proceeded slowly down the path, keeping his eyes riveted on the ground."

Dr. Watson
A Study in Scarlet
Sir Arthur Conan Doyle

Sherlock Holmes, the master of detectives, considered it essential to be excruciatingly disciplined in his approach to looking for evidence at a crime scene. While it is imperative that all investigators apply discipline in their search for evidence, it is apparent that few do so in the same way. Currently, there are no nationally accepted guidelines or standard practices for conducting explosion or bombing scene investigations. Professional training exists through Federal, State, and local agencies responsible for these investigations, as well as through some organizations and academic institutions. The authors of this *Guide* strongly encourage additional training for public safety personnel.

Purpose and Scope

The principal purpose of this *Guide* is to provide an investigative outline of the tasks that should be considered at every explosion scene. They will ensure that proper procedures are used to locate, identify, collect, and preserve valuable evidence so that it can be examined to produce the most useful and effective information—best practices. This *Guide* was designed to apply to explosion and bombing scene investigations, from highly complex and visible cases, such as the bombing of the Alfred P. Murrah Federal Building in Oklahoma City, to those that attract less attention and fewer resources but may be just as complex for the investigator. Any guide addressing investigative procedures must ensure that each contributor of evidence to the forensic laboratory system is served by the guide and that quality examinations will be rendered. Consistent collection of quality evidence in bombing cases will result in more successful investigations and prosecutions of bombing cases. While this *Guide* can be useful to agencies in developing their own procedures, the procedures included here may not be deemed applicable in every circumstance or jurisdiction, nor are they intended to be all-inclusive.

Statistics on Bombings and Other Explosives-Related Incidents

The principal Federal partners in the collection of data related to explosives incidents in the United States are the Bureau of Alcohol, Tobacco and Firearms (ATF), the Federal Bureau of Investigation (FBI), the U.S. Postal Inspection Service (USPIS), and the U.S. Fire Administration (USFA). These Federal partners collect and compile information supplied by State and local fire service and law enforcement agencies throughout the United States and many foreign countries.

According to ATF and FBI databases, there were approximately 38,362 explosives incidents from 1988 through 1997 (the latest year for which complete data were available) in the United States, including Guam, Puerto Rico, and the U.S. Virgin Islands. Incident reports received by ATF and the FBI indicate that the States with the most criminal bombing incidents are traditionally California, Florida, Illinois, Texas, and Washington. Criminal bombings and other explosives incidents have occurred in all States, however, and the problem is not limited to one geographic or demographic area of the country.

The number of criminal bombing incidents (bombings, attempted bombings, incendiary bombings, and attempted incendiary bombings) reported to ATF, the FBI, and USPIS fluctuated in the years 1993–97, ranging between 2,217 in 1997 and 3,163 in 1994. Incendiary incidents reached a high of 725 in both 1993 and 1994. Explosives incidents reached a high of 2,438 in 1994 and a low of 1,685 in 1997. It is important to note that these numbers reflect only the incidents reported to Federal databases and do not fully reflect the magnitude of the problem in the United States.

Of the criminal bombing incidents reported during 1993–97, the top three targets—collectively representing approximately 60 percent of the incidents—were residential properties, mailboxes, and vehicles. Motives are known for about 8,000 of these incidents, with vandalism and revenge by far cited most frequently.

The most common types of explosive/incendiary devices encountered by fire service and law enforcement personnel in the United States are traditionally pipe bombs, Molotov cocktails, and other improvised explosive/incendiary devices. The most common explosive materials used in these devices are flammable liquids and black and smokeless powder.

Stolen explosives also pose a significant threat to public safety in the United States. From 1993 to 1997, more than 50,000 pounds of high cxplosives, low explosives, and blasting agents and more than 30,000 detonators were reported stolen. Texas, Pennsylvania, California, Tennessee, and North Carolina led the Nation in losses, but every State reported losses.

Further information, including updated and specific statistical information, can be obtained by contacting the ATF Arson and Explosives National Repository at 800–461–8841 or 202–927–4590, through its Web site at http://ows.atf.treas.gov:9999, or by calling the FBI Bomb Data Center at 202–324–2696.

Background

National Bombing Scene Planning Panel (NBSPP)

The National Center for Forensic Science (NCFS) at the University of Central Florida (UCF) in Orlando, a grantee of the National Institute of Justice (NIJ), held a National Needs Symposium on Arson and Explosives in August 1997. The symposium's purpose was to identify problem areas associated with the collection and analysis of fire and explosion debris. One of the problem areas identified was the need for improved, consistent evidence recognition and handling procedures.

In spring 1998, NIJ and NCFS, using NIJ's template for creating technical working groups, decided to develop guidelines for fire/arson and explosion/bombing scene investigations. The NIJ Director selected members for a planning group to craft the explosion/bombing investigation guidelines—NBSPP. At the same time, the NIJ Director selected a fire/arson planning panel. The nine NBSPP members represent national and international organizations whose constituents are responsible for investigating explosion and bombing scenes and evaluating evidence from these investigations. The group also includes one academic researcher. The rationale for their involvement was twofold:

◆ They represent the diversity of the professional discipline.

◆ Each organization is a key stakeholder in the conduct of explosion and bombing investigations and the implementation of this *Guide.*

NBSPP was charged with developing an outline for national guidelines for explosion and bombing scene investigations—using the format in the NIJ publication *Death Investigation: A Guide for the Scene Investigator*[1] as a template—and identifying the expertise composition of a technical working group for explosion/bombing scene investigations. This task was completed in March 1998 at a meeting at NCFS; the results are presented here.

Technical Working Group for Bombing Scene Investigation (TWGBSI)

Candidates for TWGBSI were recommended by national law enforcement, prosecution, forensic sciences, and bomb technician organizations and commercial interests and represented a multidisciplinary group of both national and international organizations. These individuals are all content area experts who serve within the field every day. The following criteria were used to select the members of TWGBSI:

◆ Each member was nominated/selected for the position by NBSPP and NCFS.

◆ Each member had specific knowledge regarding explosion and bombing investigation.

◆ Each member had specific experience with the process of explosion and bombing investigation and the outcomes of positive and negative scene investigations.

◆ Each member could commit to the project for the entire period.

The 32 experts selected as members of TWGBSI came from 3 countries (the United States, Canada, and Israel), 13 States, and the District of Columbia. Because this technical working group dealt with explosion and bombing scenes, a large portion of investigators and analysts represented ATF and the FBI. The geographical distribution of TWGBSI members is shown in exhibit 1.

Chronology of Work

NBSPP meeting. In March 1998, the panel met at UCF, under the sponsorship of NCFS, to review the existing literature and technologies, prepare the project objectives, and begin the guideline development process. The panel's objective was to develop an outline for a set of national guidelines based on existing literature and present them for review to the assembled TWGBSI at a later date. During this initial session, five investigative tasks were identified. Each task included subsections that, when developed, provided a template of procedures for investigators to follow while conducting an explosion or bombing investigation.

Exhibit 1. Technical Working Group for Bombing Scene Investigation Membership Distribution

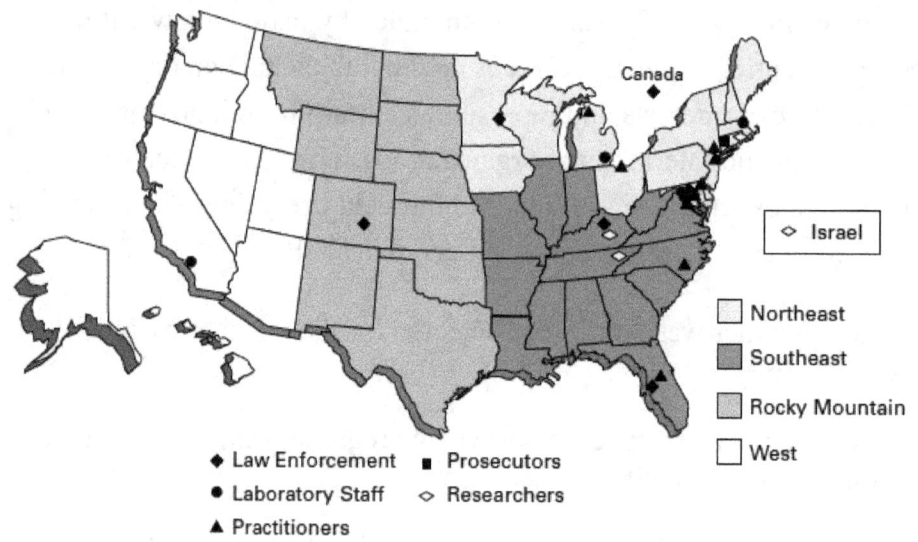

Region	Number of Participants
Northeast	20
Southeast	8
Rocky Mountain	1
West	1
International	2

The completed *Guide* includes the following components:

◆ A principle citing the rationale for performing the task.

◆ The procedure for performing the task.

◆ A summary outlining the principle and procedure.

TWGBSI assembled in August 1998. After introductory remarks from the president of UCF, TWGBSI separated into five breakout sections to draft the *Guide,* which includes the following stages:

◆ Prioritizing initial response efforts.

◆ Evaluating the scene.

◆ Documenting the scene.

◆ Processing evidence at the scene.

◆ Completing and recording the scene investigation.

Once all breakout groups completed their work, the full group reassembled to review and approve the initial draft. Editors from an NIJ contractor attended each section to record the proceedings and guide the editorial process. After the meeting, the editors reformatted the initial draft and forwarded it to an agency representative so that it could be sent to all TWGBSI members for comment.

Organizational review and national reviewer network. After the TWGBSI comments were received by NIJ, NBSPP met in November and December 1998 in Washington, D.C., to consider and incorporate the comments, creating the second draft of the *Guide*. In addition, NBSPP members recommended organizations, agencies, and individuals they felt should comment on the draft document, which was mailed to all TWGBSI members and to this wider audience in June 1999. The 150 organizations and individuals whose comments were solicited during the national review of this *Guide* included all levels of law enforcement, regional and national organizations, and bomb response units from the United States, Canada, and other nations. A list of reviewers can be found in appendix C.

NBSPP members reassembled in August 1999 to incorporate the comments received from the initial wide review. Following this meeting, a third draft of the *Guide* was sent to all TWGBSI members for discussion and review within their organizations and agencies. In October 1999, the TWGBSI members met to review and recommend changes to this third draft. Another national and organizational review followed, and results were discussed by TWGBSI at a meeting in January 2000. What follows is the final consensus document resulting from the final meeting.

Training

For each of the procedures presented in this *Guide,* training criteria will be developed and approved by NCFS's Technical Working Group on Fire and Explosions. These criteria will provide individuals and educational organizations with an additional resource for providing comprehensive instruction to public safety personnel. A current listing of institutions that can provide training in the area of explosion/bombing investigation can be obtained from NCFS (see appendix D).

Authorization

Federal, State, and local statutory authority in explosion and bombing cases is enforced by the agencies responsible for the specific incident and varies greatly depending on the specific location and nature of the incident.

Note

1. *Death Investigation: A Guide for the Scene Investigator,* Research Report, Washington, D.C.: U.S. Department of Justice, National Institute of Justice, December 1997, NCJ 167568.

A Guide for Explosion and Bombing Scene Investigation

Section A	Procuring Equipment and Tools
Section B	Prioritizing Initial Response Efforts
Section C	Evaluating the Scene
Section D	Documenting the Scene
Section E	Processing Evidence at the Scene
Section F	Completing and Recording the Scene Investigation

Section A. Procuring Equipment and Tools

Possessing the proper tools and equipment is key to any task, and never more so than in emergency situations such as explosion or bombing scenes. Because responders and investigators may not know the details of the situation until arriving at the scene, prior preparation is vital. Following is a list of equipment and tools frequently used by the investigative team at explosion and bombing scenes. Equipment and tool needs are, for the most part, determined by the actual scene. The list below may be used as a planning guide for equipment and tool needs. Not every item and tool mentioned below will be applicable for use on every scene.

Safety

◆ Biohazard materials (i.e., bags, tags, labels).

◆ First-aid kit.

◆ Footwear, safety (i.e., protective shoes/boots).

◆ Glasses, safety.

◆ Gloves, heavy and disposable (e.g., surgical, latex).

◆ Helmets, safety/hard hats.

◆ Kneepads.

◆ Outerwear, protective (e.g., disposable suits, weather gear).

◆ Personnel support items (e.g., food, water, hygiene items, shelter).

◆ Reflective tape.

◆ Respiratory equipment (e.g., particle masks, breathing equipment).

General Crime Scene Tools/Equipment

◆ Barrier tape/perimeter rope.

◆ Batteries.

◆ Binoculars.

◆ Communications equipment (e.g., telephone, two-way radio).

◆ Evidence collection kits (e.g., latent print, bodily fluid, impression, tool mark, trace evidence).

◆ Flares.

◆ Flashlights.

◆ Generators.

◆ Handtools (e.g., screwdrivers, crowbars, hammers).

◆ Knives, utility.

◆ Lighting, auxiliary.

◆ Tarps/tents.

◆ Thermometer.

◆ Trashcans, large.

◆ Tweezers/forceps.

Scene Documentation

◆ Compass.

◆ Computer and computer-aided design (CAD) program.

◆ Consent-to-search forms.

◆ Drawing equipment (e.g., sketchbooks, pencils).

◆ Logs (e.g., evidence recovery, photo).

- Measuring equipment (e.g., forensic mapping station, tape measure, tape wheel).

- Photographic equipment (e.g., 35mm camera, Polaroid camera, videocamera, digital camera, film, lenses, tripods).

- Tape recorder and cassettes.

- Writing equipment (e.g., notebooks, pens, permanent markers).

Evidence Collection

- Bags, new (e.g., sealable, nylon).

- Boxes, corrugated/fiberboard.

- Brushes and brooms.

- Cans, new (e.g., unlined).

- Evidence flags/cones.

- Evidence placards.

- Evidence tags.

- Evidence sealing tape.

- Gloves (i.e., disposable cotton, disposable latex).

- Grid markers.

- Heat sealer.

- Magnets.

- Outerwear, protective (e.g., disposable suits, shoe covers).

- Rakes, spades, and shovels.

- Sifters/screens.

- Swabbing kits.

- Trowels.

- Vacuum.

Specialized Equipment

◆ Aerial survey/photography equipment (e.g., helicopter).

◆ Chemical test kits and vapor detectors.

◆ Construction equipment, heavy.

◆ Extrication/recovery equipment.

◆ GPS (global positioning system) equipment.

◆ Ladders.

◆ Trace explosives detectors (e.g., sniffers) and/or detection canines.

Note: Safety concerns should be continually addressed beginning with the initial response effort. Implementation of the procedures in this section will be determined by the scene circumstances.

1. Conduct a Preliminary Evaluation of the Scene

B

Principle: First responders (the first public safety personnel to arrive at the scene, whether law enforcement officers, firefighters, or emergency medical services (EMS) personnel) must assess the scene quickly yet thoroughly to determine the course of action to be taken. This assessment should include the scope of the incident, emergency services required, safety concerns, and evidentiary considerations.

Procedure: Upon arrival at the scene, first responders should:

A. Establish a command post/implement an incident command system (i.e., a point of contact and line of communication and authority for other public safety personnel).

B. Request emergency services from bomb technicians, firefighters, EMS personnel, and law enforcement officers.

C. Identify scene hazards, such as structural collapse, blood-borne pathogens, hazardous chemicals, and secondary explosive devices.

D. Identify witnesses, victims, and the presence of evidence.

E. Preserve potentially transient physical evidence (e.g., evidence present on victims, evidence that may be compromised by weather conditions).

> **DANGER: Beware of secondary devices!**
> The scene may contain secondary explosive devices designed specifically to kill or maim public safety responders. Do not touch any suspicious items. If a suspected secondary device is located, immediately evacuate the area and contact bomb disposal personnel.

Summary: Based on the preliminary evaluation, first responders will initiate an incident command system, request emergency services, and identify scene hazards and evidentiary concerns.

2. Exercise Scene Safety

Principle: Safety overrides all other concerns. First responders must take steps to identify and remove or mitigate safety hazards that may further threaten victims, bystanders, and public safety personnel. They must exercise due caution while performing emergency operations to avoid injuries to themselves and others.

Procedure: Following the preliminary evaluation of the scene, first responders should:

A. Request additional resources and personnel (e.g., bomb technicians, building inspectors, representatives from utility companies, such as gas, water, and electric) to mitigate identified hazards.

B. Use tools and personal protective equipment appropriate to the task during all operations.

C. Request and/or conduct a safety sweep of the area by personnel qualified to identify and evaluate additional hazards and safety concerns.

D. Mark hazard areas clearly and designate safety zones to receive victims and evacuees.

Summary: To ensure safety, first responders will take steps to identify, evaluate, and mitigate scene hazards and establish safety zones.

3.   Administer Lifesaving Efforts

Principle: First responders' primary responsibility is to rescue living victims and provide treatment for life-threatening injuries. While performing emergency operations, they are to preserve evidence and avoid disturbing areas not directly involved in the rescue activities, including those areas containing fatalities.

Procedure: After performing a preliminary evaluation and establishing scene safety, first responders should:

A. Initiate rescues of severely injured and/or trapped victims.

B. Evacuate ambulatory victims, perform triage, and treat life-threatening injuries.

C. Leave fatalities and their surroundings undisturbed. Removal of fatalities will await authorization.

D. Avoid disturbing areas not directly involved in rescue activities.

Summary: Lifesaving efforts are first responders' priority. Additionally, care should be taken not to disturb areas where rescue activities are not taking place.

4.   Establish Security and Control

Principle: First responders will establish control and restrict scene access to essential personnel, thereby aiding rescue efforts and scene preservation. First responders will initiate documentation.

Procedure: To establish security and control, first responders should:

    A. Set up a security perimeter.

    B. Restrict access into and out of the scene through the security perimeter (e.g., control media, bystanders, nonessential personnel).

    C. Establish staging areas to ensure that emergency vehicles have access into the area.

    D. Initiate documentation of the scene as soon as conditions permit (e.g., taking notes, identifying witnesses, videotaping/photographing bystanders).

Summary: First responders will establish a controlled security perimeter, designate staging areas, and initiate documentation. This will set the stage for the subsequent investigation.

Section C. Evaluating the Scene

Note: At the time the scene is determined to involve a bombing or other crime, the investigator must address legal requirements for scene access, search, and evidence seizure.

1. Define the Investigator Role

Principle: The investigator must coordinate with the incident commander and first responders to determine what occurred and to assess the current situation. Subsequent procedures will vary depending on the magnitude of the incident.

Procedure: Upon arriving at and prior to entering the scene, the investigator should:

A. Identify and introduce himself or herself to the incident commander.

B. Interview the incident commander and first responders to evaluate the situation, including safety concerns, and determine the level of investigative assistance needed.

C. Conduct a briefing with essential personnel (e.g., law enforcement, fire, EMS, hazardous materials, and utility services personnel) to:

◆ Evaluate initial scene safety to the extent possible prior to entry.

◆ Ensure that a search for secondary explosive devices has been conducted.

Caution: Only bomb disposal personnel should handle any suspected devices that are located. Take no further action until the devices have been identified or rendered safe.

- Ensure that the scene has been secured, that a perimeter and staging areas for the investigation have been established, and that all personnel have been advised of the need to prevent contamination of the scene.

- Ensure that the chain of custody is initiated for evidence that may have been previously collected.

D. Assess legal considerations for scene access (e.g., exigent circumstances, consent, administrative/criminal search warrants).

Summary: The investigator will conduct a briefing to ensure scene safety and security, while addressing the issue of secondary devices.

2. Ensure Scene Integrity

Principle: The investigator must ensure the integrity of the scene by establishing security perimeters and staging areas, contamination control procedures, and evidence collection and control procedures.

Procedure: Prior to evidence collection, the investigator should:

A. Establish procedures to document personnel entering and exiting the scene.

B. Establish and document procedures to prevent scene contamination.

C. Establish and document procedures for evidence collection, control, and chain of custody (see the sample evidence recovery and chain of custody logs in appendix A).

Summary: The investigator will establish and document procedures to protect the integrity of the scene.

3.   Conduct the Scene Walkthrough

Principle: The investigator must conduct a walkthrough to establish scene parameters and acquire an overview of the incident.

Procedure: During the scene walkthrough, the investigator should:

A. Reevaluate scene requirements (e.g., boundaries, personnel, equipment).

B. Establish an entry and exit path for personnel.

C. Be alert to safety concerns (e.g., structural damage, secondary devices, unconsumed explosive materials, failed utilities, hazardous materials) and to the locations of physical evidence.

D. Ensure preservation and/or collection of transient evidence.

E. Attempt to locate the seat(s) of the explosion(s).

Summary: The investigator's initial walkthrough will be an opportunity to identify evidence and the presence of safety hazards.

4.   Secure Required Resources

Principle: Following the walkthrough, the investigator should meet with available emergency responders and investigative personnel to determine what resources, equipment, and additional personnel may be needed.

Procedure: During the course of this meeting, the investigator should:

A. Assess the nature and scope of the investigation through information obtained during the walkthrough and from all available personnel.

B. Advise personnel of any secondary devices or other hazards found at the scene.

C. Ensure that one list of victims/potential witnesses is developed and that their accounts of the incident are documented.

D. Ensure that required evidence collection equipment, as well as processing and storage facilities, are available.

E. Secure required equipment as determined by the scene conditions, such as light and heavy equipment, handtools, specialty equipment, and personal safety items.

F. Ensure that sufficient utilities and support services are requested (e.g., electricity, food, trash removal, sanitary services, other public services, security).

G. Advise emergency responders and the investigation team of their assignments for scene documentation and processing.

H. Remind personnel that evidence can take many forms; it is not limited solely to components of the device(s).

Summary: The investigator will meet with emergency responders and investigative personnel in preparation for scene documentation and processing.

Section D. Documenting the Scene

1. Develop Written Documentation

Principle: The investigator will prepare written scene documentation to become part of the permanent record.

Procedure: The investigator should:

A. Document access to the scene (see the sample access control log in appendix A).

B. Document activities, noting dates and times, associated with the incident and the investigation (see the sample activity log in appendix A).

C. Describe the overall scene in writing, noting physical and environmental conditions (e.g., odors, weather, structural conditions) (see the sample narrative description in appendix A).

D. Diagram and label scene features using sketches, floor plans, and architectural or engineering drawings.

E. Describe and document the scene with measuring equipment, which may include surveying equipment, GPS (global positioning system) technology, or other available equipment.

Summary: Investigators must prepare written scene documentation as part of the permanent record of the incident, which will serve as the foundation for any incident reconstructions and future proceedings.

2. Photograph/Videotape the Scene

Principle: The investigator must ensure that photographic documentation is included in the permanent scene record. This documentation should be completed prior to the removal or disturbance of any items.

Procedure: The investigator should:

A. Record overall views of the scene (e.g., wide angle, aerial, 360-degree) to spatially relate items within and to the scene and surrounding area. (A combination of still photography, video-taping, and other techniques is most effective.)

B. Consider muting the audio portion of any video recording unless there is narration.

C. Minimize the presence of scene personnel in photographs/videos.

D. Consider photographing/videotaping the assembled crowd.

E. Maintain photo and video logs (see the sample photographic log in appendix A).

Summary: The investigator will ensure the photographic documentation of the scene to supplement the written documentation in preparation for scene reconstruction efforts and any future proceedings.

3.   Locate and Interview Victims and Witnesses

Principle: The investigator will obtain victims'/witnesses' identities, statements, and information concerning their injuries.

Procedure: The investigator should:

A. Identify and locate witnesses (e.g., victims who may have been transported, employees, first responders, delivery/service personnel, neighbors, passers-by) and prioritize interviews.

B. Attempt to obtain all available identifying data regarding victims/witnesses (e.g., full name, address, date of birth, work and home telephone numbers) prior to their departure from the scene.

C. Establish each witness' relationship to or association with the scene and/or victims.

D. Establish the basis of the witness' knowledge: How does the witness have knowledge of the incident?

E. Obtain statements from each witness.

F. Document thoroughly victims' injuries and correlate victims' locations at the time of the incident with the seat(s) of the explosion(s).

G. Interview the medical examiner/coroner and hospital emergency personnel regarding fatalities and injuries.

Summary: The investigator must attempt to determine the locations of all victims and witnesses. Victim and witness statements and information about their injuries may be essential to establishing the nature of the device and the circumstances of the incident.

Section E. Processing Evidence at the Scene

> **Note:** *At the time the scene is determined to involve a bombing or other crime, the investigator must address legal requirements for scene access, search, and evidence seizure.*

1. Assemble the Evidence Processing Team

Principle: Effective organization and composition of the evidence processing team ensure the proper collection and preservation of evidence.

Procedure: The size of the evidence processing team depends on the magnitude of the scene, but the investigator needs to ensure that the following roles and expertise are addressed:

A. Bomb disposal technician.

B. Evidence custodian.

C. Forensic specialist.

D. Logistics specialist.

E. Medical examiner.

F. Photographer (still, digital, video, etc.).

G. Procurement specialist.

H. Safety specialist (structural engineer, etc.).

I. Searchers/collectors.

J. Sketch artist.

Summary: Attention to the organization and composition of the evidence processing team facilitates effective evidence collection and preservation.

E

2.   Organize Evidence Processing

Principle: Good organization is essential to evidence collection and preservation. The investigator must continually evaluate the scene, adapt to changes as they occur, and brief the team.

Procedure: Before deploying the team, the investigator should:

A. Review and reevaluate:

- ◆ The boundaries of the scene.
- ◆ Safety concerns.
- ◆ Command post and staging locations.
- ◆ Evidence processing and storage locations.
- ◆ Personnel and equipment requirements.
- ◆ Legal and administrative considerations.

B. Identify the search procedure for the scene.

C. Ensure that transient physical evidence has been preserved and collected.

D. Consider onsite explosives detection (e.g., trace explosives detection, use of canines, chemical tests) by qualified personnel.

E. Brief the team and review assignments.

Summary: Prior to evidence collection and throughout the process, the investigator will review the scene, adapt to changes, and brief the team.

3.   Control Contamination

Principle: Preventing contamination protects the integrity of the scene and other search areas, the integrity of the evidence for forensic analyses, and the safety of personnel.

28

Procedure: The investigator should ensure that evidence processing personnel:

A. Use clean protective outergarments and equipment as applicable for each scene.

B. Consider obtaining control samples as applicable (e.g., evidence containers, swabs of equipment and personnel).

C. Package collected evidence in a manner that prevents loss, degradation, or contamination.

D. Package, store, and transport evidence from different scenes or searches in separate external containers.

Summary: Proper collection, packaging, transportation, and storage will minimize contamination and ensure the integrity of the evidence.

4. Identify, Collect, Preserve, Inventory, Package, and Transport Evidence

Principle: The search focuses on the discovery of physical evidence that may establish that a crime was committed and link elements of the crime to possible suspects.

Procedure: To maximize the recovery and evaluation of all types of physical evidence, the investigator should ensure:

A. The preparation of an evidence recovery log (see the sample in appendix A) that documents information such as:

- ◆ Item number.
- ◆ Description.
- ◆ Location found (grid number if used).
- ◆ Collector's name.
- ◆ Markings (either directly on the item or indirectly on the package).

- ◆ Packaging method.
- ◆ Miscellaneous comments.

B. The identification of evidence by:

- ◆ Assigning personnel to designated search areas.
- ◆ Initiating scene-specific search pattern(s) and procedures, including examination of immobile structures for possible evidence.
- ◆ Attempting to determine the method of bomb delivery.
- ◆ Establishing the seat(s) of the explosion(s), if present.
- ◆ Documenting blast effects (e.g., structural damage, bent signs, thermal effects, fragmentation).
- ◆ Examining the crater, vehicles, structures, etc.
- ◆ Documenting the location(s) of victims prior to and after the explosion.
- ◆ Ensuring that victims are examined for bomb component fragments. Autopsies should include full-body x-rays.

C. The collection of evidence, including:

- ◆ Suspected bomb components and fragments, including those recovered from victims.
- ◆ Suspected materials used in the construction and transportation of the explosive device(s) (e.g., tape, batteries, manuals, vehicles).
- ◆ Crater material.
- ◆ Residues and other trace evidence (using swabbing techniques).
- ◆ Additional items of evidence (e.g., blood, hair, fiber, fingerprints, tire tracks, weapons, documents, tools).
- ◆ Comparison samples of indigenous materials.

D. That evidence is:

- ◆ Photographed.
- ◆ Packaged and preserved in containers.
- ◆ Labeled (e.g., date, collector's initials, item number, location).
- ◆ Recorded in the evidence recovery log.
- ◆ Secured in the designated storage location.

E. The labeling, transportation, and storage of evidence by:

- ◆ Placing evidence from different locations or searches in separate external containers.

- ◆ Labeling evidence for storage and shipment, including identification of hazards.

- ◆ Arranging for transportation of the evidence.

Summary: Identification, collection, preservation, and packaging of evidence must be conducted in a manner that protects the item, minimizes contamination, and maintains the chain of custody. These steps assist in establishing the elements of a possible crime and provide the basis for thorough, accurate, and objective investigation and prosecution processes.

Section F. Completing and Recording the Scene Investigation

1. Ensure That All Investigative Steps Are Documented

Principle: To ensure that the permanent record will be complete, the investigator should review all documentation before releasing the scene.

Procedure: The investigator should verify that the following have been addressed:

A. Documentation of major events and time lines related to the incident.

B. Personnel access log (see the sample in appendix A).

C. Activity log (see the sample in appendix A).

D. Review of interviews and events.

E. Narrative description of the scene (see the sample in appendix A).

F. Photo and video logs (see the sample in appendix A).

G. Diagrams, sketches, and evidence mapping.

H. Evidence recovery log (see the sample in appendix A).

Summary: By accounting for all investigative steps prior to leaving the scene, the investigator ensures an accurate and thorough representation of the scene for the permanent record.

33

2. Ensure That Scene Processing Is Complete

Principle: The scene may be released only upon conclusion of the onsite investigation and a thorough evidence collection process.

Procedure: The investigator should perform a critical review of the scene investigation with all personnel, to include the following actions:

A. Discuss with team members, including those not present at the scene, preliminary scene findings and critical issues that arose during the incident.

B. Ensure that all identified evidence is in custody.

C. Recover and inventory equipment.

D. Decontaminate equipment and personnel.

E. Photograph and/or videotape the final condition of the scene just before it is released.

F. Address legal considerations.

G. Discuss postscene issues (e.g., forensic testing, insurance inquiries, interview results, criminal histories).

H. Communicate and document postscene responsibilities.

Summary: The investigator will review the scene investigation to ensure that it is complete and that postscene issues are addressed.

3. Release the Scene

Principle: The release of the scene must be documented. The investigator should ensure communication of known scene-related health and safety issues to a receiving authority at the time of release.

Procedure: Upon releasing the scene, the investigator should:

A. Address public health and safety issues by performing the following tasks:

◆ Contacting public utilities.

◆ Evaluating biological and chemical hazards.

◆ Evaluating structural integrity issues.

◆ Assessing environmental issues.

B. Identify a receiving authority for the scene.

C. Ensure disclosure of all known health and safety issues to a receiving authority.

D. Document the time and date of release, to whom the scene is being released, and by whom.

Summary: The investigator will ensure communication of known health and safety issues to a receiving authority upon releasing the scene and will document the release.

4. Submit Reports to the Appropriate National Databases

Principle: Detailed technical information regarding explosive devices is collected, integrated, and disseminated via national databases. These data help authorities identify the existence of serial bombers, the sophistication of explosive devices being used, and the need for uniform procedures and further development of equipment.

Procedure: The investigator or authorized agency's administration should submit detailed reports to these databases:

A. Arson and Explosives National Repository (Bureau of Alcohol, Tobacco and Firearms).

B. Bomb Data Center (Federal Bureau of Investigation).

C. Uniform Crime Reports, National Incident-Based Reporting System, and National Fire Incident Reporting System.

Summary: The investigator contributes to the compilation of national databases that identify trends in explosions and other incidents involving explosives.

Appendix A. Sample Forms

Following are sample forms that can be adapted for use as needed.

A1.	Consent to Search

A2.	Access Control Log

A3.	Activity Log

A4.	Narrative Description

A5.	Photographic Log

A6.	Evidence Recovery Log

A7.	Evidence Control/Chain of Custody

CONSENT TO SEARCH

I, _____
         (Person giving consent)

having been informed of my constitutional right not to have a search made of my premises without a search warrant and of my right to refuse to consent to such a search, do authorize Fire or Police Investigator,

_____
         (Name of Investigator)

or his designee, to conduct a complete search of my premises known as

_____
         (Address of property)

for the purpose of establishing the cause of the explosion which occurred at my premises on_____ .
         (Date of explosion)

I am aware that the search is being conducted to search for evidence of the cause of the explosion and I agree to allow the above-named investigator or his designee to take photographs/videotapes of the premises, to remove papers, letters, materials, or other property, knowing that they may be submitted for forensic examination and testing.

I am aware that the above-named investigator or his designee will be on the premises for a period of time and I have no objection to their entering and remaining on the premises for a number of days. This written consent is being given by me voluntarily and without threats or promises of any kind.

I know that I can refuse to give this consent to search and I am waiving that right signing this consent.

                                   _____
                                            Person Giving Consent

_____
Witness

_____
Witness

38

ACCESS CONTROL LOG

Date:_____ Starting Time:_____ Platoon:_____ OCA or Dispatch #:_____

Type of Crime:

Location of Crime:

Name	Position/Title	Time In	Time Out

Remarks:

Initiated By:_____ Initiating Officer:_____
            Print Name/Call Sign   Signature

Relieved By:_____ Relieving Officer:_____
            Print Name/Call Sign   Signature

Date/Time Relieved:_____ _____ Date/Time Completed:_____ _____
                    Date   Time   Date   Time

Supervisor's Signature:_____

The completed form is to be turned over to the Investigating Detective   Page___ of ___

ACTIVITY LOG

LOCATION

CASE IDENTIFIER

PREPARER/ASSISTANTS

TIME AND DATE OF ARRIVAL

PERSON(S) PRESENT AT SCENE
AT TIME OF ARRIVAL

PERSONNEL INVOLVED
IN SEARCH AND THEIR
RESPECTIVE DUTIES

PRELIMINARY SURVEY/EVIDENCE EVALUATION (NOTES/OBSERVATIONS)

PERSON IN CHARGE OF SCENE
AT TIME OF ARRIVAL

CONTROL OF SCENE OBTAINED
FROM

TIME AND DATE OF CONTROL
ACQUISITION

CONDITION OF SCENE
ON ARRIVAL

(SECURED/UNSECURED)

LIGHTING CONDITIONS

WEATHER CONDITIONS

SCENE CONSISTS OF

(GENERAL DESCRIPTION)

SPECIAL SITUATIONS OR CONDITIONS (NOTES/OBSERVATIONS)

(Continued)

40

ACTIVITY LOG (Continued)

CRIME SCENE CONFERENCE (NOTES/OBSERVATIONS)	CRIME SCENE INVESTIGATION COMPLETED and RELEASE OF SCENE AUTHORIZED	NAME
		TITLE
		DATE
		TIME
		SIGNATURE
	CRIME SCENE RELEASED TO	NAME
		TITLE
		DATE
		TIME
	(ADDITIONAL NOTES/OBSERVATIONS)	

FINAL SURVEY (NOTES/OBSERVATIONS)

ADMINISTRATIVE LOG:

TIME	PERTINENT DESCRIPTION / INFORMATION

NARRATIVE DESCRIPTION

LOCATION _____
DATE _____
CASE IDENTIFIER _____
PREPARER/ASSISTANTS _____

MODE of NARRATIVE PREPARATION _____

42

PHOTOGRAPHIC LOG

LOCATION

DATE

CASE IDENTIFIER

PREPARER/ASSISTANTS

CAMERA

TYPE of FILM and RATING

REMARKS

PHOTO #	DESCRIPTION of PHOTOGRAPHIC SUBJECT	USE of SCALE	MISCELLANEOUS COMMENTS

SKETCH (IF APPLICABLE)

EVIDENCE RECOVERY LOG

LOCATION_____
DATE_____
CASE IDENTIFIER_____
PREPARER/ASSISTANTS_____

PERSONNEL_____

ITEM #	DESCRIPTION	WHERE FOUND	RECOVERED BY	PHOTO	MARKING Direct—O Indirect—I	PACKAGING METHOD	MISCELLANEOUS COMMENTS

EVIDENCE CONTROL/CHAIN OF CUSTODY

Please Furnish Complete Information

Agency submitting evidence		Date
		Laboratory #

Delivered by	Accepted by	

Suspect(s)	Victim(s)

Offense	Place and date of offense
Agency case #	

Date of hearing, grand jury, trial, or reason why expeditious handling is necessary

Prev. exams this case ☐ Yes ☐ No	Evid. located Room # _____	Report to be directed to
Copies to		Evidence to be returned to ☐ Mailed Back ☐ Picked Up by Contributor

Brief Facts covering case

Description of evidence	Exams requested
	(This space for blocking)

45

Item	Received by	Date	Time	Received by	Date	Time
	Signature_____ Reason _____			Signature_____ Reason _____		
	Signature_____ Reason _____			Signature_____ Reason _____		
	Signature_____ Reason _____			Signature_____ Reason _____		
	Signature_____ Reason _____			Signature_____ Reason _____		
	Signature_____ Reason _____			Signature_____ Reason _____		
	Signature_____ Reason _____			Signature_____ Reason _____		
	Signature_____ Reason _____			Signature_____ Reason _____		
	Signature_____ Reason _____			Signature_____ Reason _____		

REMARKS _____
_____
_____
_____
_____
_____

Appendix B. Further Reading

Beveridge, A. *Forensic Investigation of Explosions*. London: Taylor & Francis Ltd., 1998.

Conkling, J.A. *Chemistry of Pyrotechnics and Explosives*. New York: Marcel Dekker, Inc., 1985.

Cook, M.A. *The Science of High Explosives*. Malabar, Florida: Robert E. Krieger Publishing Company, 1958, 1985.

Cooper, P.W. *Explosives Engineering*. New York: Wiley-VCH, 1997.

Cooper, P.W., and S.R. Kurowski. *Introduction to the Technology of Explosives*. New York: Wiley-VCH, 1997.

Davis, T.L. *The Chemistry of Powder and Explosives*. Hollywood, California: Angriff Press, 1972.

DeHaan, J.D. *Kirk's Fire Investigation*. 4th ed. Indianapolis: Brady Publishing/Prentice Hall, 1997.

Encyclopedia of Explosives and Related Items. Vols. 1–10. Dover, New Jersey: Picatinny Arsenal, U.S. Army Armament Research and Development Command, 1960–83.

Kennedy, P.M., and J. Kennedy. *Explosion Investigation and Analysis: Kennedy on Explosions*. Chicago: Investigations Institute, 1990.

The ISEE Blaster's Handbook. 17th ed. Cleveland: International Society of Explosives Engineers, 1998.

Kohler, J., and R. Meyer. *Explosives*. 4th, revised and extended ed. New York: Wiley-VCH, 1993.

Military Explosives. U.S. Army and U.S. Air Force Technical Manual TM 9–1300–214. Washington, D.C.: U.S. Army, 1967.

National Fire Protection Association. *NFPA 921: Guide for Fire and Explosion Investigations.* Quincy, Massachusetts: National Fire Protection Association.

Urbanski, T. *Chemistry and Technology of Explosives.* Vols. 1–4. New York: Pergamon Press, 1983.

Yinon, J., and S. Zitrin. *Modern Methods and Applications in Analysis of Explosives.* New York: Wiley-VCH, 1993.

Appendix C. List of Organizations

During the review process, drafts of this document were sent to the following agencies and organizations for comment. While TWGBSI considered all comments and issues raised by these organizations, this *Guide* reflects only the positions of its authors. Mention of the reviewers is not intended to imply their endorsement.

Accomack County (VA) Sheriff's Office

Alaska State Criminal Laboratory

American Academy of Forensic Sciences

American Bar Association

American Correctional Association

American Jail Association

American Prosecutors Research Institute

American Reinsurance Company

American Society of Crime Laboratory
  Directors

American Society of Law Enforcement Trainers

Anchorage (AK) Police Department

Arapahoe County (CO) Sheriff's Office

Armstrong Forensic Laboratory

Association of Federal Defense Attorneys

Bridgeport (MI) Forensic Laboratory

Bristol (VA) Police Department

Broward County (FL) Sheriff's Office

Brownsville (TX) Police Department

Bureau of Alcohol, Tobacco and Firearms

Cameron County (TX) Sheriff's Office

Campaign for Effective Crime Policy

Chicago (IL) Fire Department

Cincinnati (OH) Fire Division

City of Donna (TX) Police Department

City of Inver Grove Heights (MN) Fire Marshal

Clark County (NV) Fire Department

Cleveland State College Basic Police Academy

Commission on Accreditation of Law
  Enforcement Agencies

Conference of State Court Administrators

Connecticut State Police Forensic Laboratory

Conyers (GA) Police Department

Council of State Governments

Covington (TN) Fire Department

Criminal Justice Institute

Delaware State Fire Marshal's Office

Drug Enforcement Administration

Edinburg (TX) Police Department

Fairbanks (AK) Police Department

Federal Bureau of Investigation

Federal Law Enforcement Training Center, U.S.
  Department of the Treasury

Florida Department of Law Enforcement

Florida State Fire Marshal

Georgia Bureau of Investigation

Georgia Public Safety Training Center

Town of Goshen (NY) Police Department

Harlingen (TX) Police Department

Hidalgo County (TX) Sheriff's Office

Illinois State Police

Indiana State Police Laboratory

Institute of Police Technology and Management

International Association for Identification

International Association of Bomb Technicians
  and Investigators

International Association of Chiefs of Police

International City/County Management
  Association

Iowa Division of Criminal Investigation
  Laboratory

Jefferson Parish (LA) Fire Department

Juneau (AK) Police Department

Laredo (TX) Police Department

Law Enforcement Training Institute

Los Angeles (CA) Fire Department

Maine State Police Crime Laboratory

Massachusetts State Fire Marshal's Office

Massachusetts State Police Crime Laboratory

McAllen (TX) Police Department

Metro Nashville (TN) Police Department

Michigan Department of State Police

Mission (TX) Police Department

National Association of Attorneys General

National Association of Black Women Attorneys

National Association of Counties

National Association of Criminal Defense
Lawyers

National Association of Drug Court Professionals

National Association of Police Organizations, Inc.

National Association of Sentencing Commissions

National Association of State Alcohol and Drug
Abuse Directors

National Association of Women Judges

National Black Police Association

National Center for State Courts

National Conference of State Legislatures

National Council on Crime and Delinquency

National Crime Prevention Council

National Criminal Justice Association

National District Attorneys Association

National Governors Association

National Institute of Standards and Technology,
Office of Law Enforcement Standards

National Law Enforcement and Corrections
Technology Centers

National Law Enforcement Council

National League of Cities

National Legal Aid and Defender Association

National Organization of Black Law Enforcement
Executives

National Sheriffs' Association

New Hampshire State Police Forensic
Laboratory

New Jersey State Police

New York State Office of Fire Prevention and
Control

Orange County (CA) Sheriff's Department

Pan American Police Department (Edinburg, TX)

Peace Officer's Standards and Training

Pennsylvania State Police Laboratory

Pharr (TX) Police Department

Pinellas County (FL) Forensic Laboratory

Police Executive Research Forum

Police Foundation

Port Authority of NY & NJ Police

Rhode Island State Crime Laboratory

St. Louis (MO) Metropolitan Police Department

San Diego (CA) Police Department

Sitka (AK) Police Department

South Carolina Law Enforcement Division

Suffolk County (NY) Crime Laboratory

Tennessee Bureau of Investigation

Tennessee Law Enforcement Training Academy

Texas Rangers Department of Public Safety

Tucson (AZ) Police Department

U.S. Border Patrol

U.S. Conference of Mayors

Utah State Crime Scene Academy

Webb County (TX) Sheriff's Department

Weslaco (TX) Police Department

Willacy County (TX) Sheriff's Office

Wisconsin State Crime Laboratory

Appendix D. Investigative and Technical Resources

Depending on jurisdiction, an investigator's initial points of contact should be his or her State or local fire marshal's office and/or State police. Following is a list of resources that may be useful for further investigative or technical information and assistance. This list should not be considered all-inclusive.

A current list of institutions that can provide training in explosion/bombing scene investigation is available from the National Center for Forensic Science.

Bureau of Alcohol, Tobacco and Firearms*
Headquarters Enforcement Operations Center
888–ATF–BOMB
202–927–8050
URL: http://www.atf.treas.gov

Arson and Explosives National Repository
800–461–8841
202–927–4590

Arson and Explosives Programs Division
202–927–7930

National Laboratory
301–762–9800

Chemical Transportation Emergency Center (CHEMTREC®)
800–262–8200
URL: http://www.chemtrec.org

Environmental Protection Agency
National Response Center
800–424–8802
URL: http://www.epa.gov

Federal Bureau of Investigation*
Bomb Data Center (preblast issues)
202–324–2696
URL: http://www.fbi.gov

Explosives Unit (postblast issues)
202–324–4341

Federal Emergency Management Agency
U.S. Fire Administration
202–447–1000
URL: http://www.usfa.fema.gov

Institute of Makers of Explosives
202–429–9280
URL: http://www.ime.org

51

International Association of Arson Investigators

314–739–4224

URL: http://www.fire-investigators.org

International Association of Bomb Technicians and Investigators

941–353–6843

URL: http://www.iabti.org

International Society of Explosives Engineers

440–349–4004

URL: http://www.isee.org

National Center for Forensic Science

407–823–6469

URL: http://www.ncfs.ucf.edu

National Fire Protection Association

617–770–3000

URL: http://www.nfpa.org

National Institute of Standards and Technology

Building and Fire Research Laboratory

301–975–6850

URL: http://www.bfrl.nist.gov

Royal Canadian Mounted Police*

Canadian Bomb Data Center

613–993–7880

URL: http://www.rcmp-grc.gc.ca/

U.S. Chemical Safety and Hazard Investigation Board

202–261–7600

URL: http://www.chemsafety.gov

U.S. Postal Inspection Service*

Forensic/Technical Services Division

703–406–7100

URL: http://www.usps.gov/websites/depart/inspect/

* These are sources for criminal investigative assistance.